Jose Antonio Moreno Serrano
Wagner Roberto Morocho Chamba
Luis Antonio Jimenez Ávila

Application of X-Ray Cryo-Tomography Technique (Cryo-XT)

Jose Antonio Moreno Serrano
Wagner Roberto Morocho Chamba
Luis Antonio Jimenez Ávila

Application of X-Ray Cryo-Tomography Technique (Cryo-XT)

Study of Viral Factories in cells infected with vaccinia virus

ScienciaScripts

This book is a translation from the original published under ISBN 978-613-9-06450-2.

Publisher:
Sciencia Scripts
is a trademark of
Dodo Books Indian Ocean Ltd. and OmniScriptum S.R.L publishing group

120 High Road, East Finchley, London, N2 9ED, United Kingdom
Str. Armeneasca 28/1, office 1, Chisinau MD-2012, Republic of Moldova, Europe
Printed at: see last page
ISBN: 978-620-7-70115-5

APPLICATION OF THE TECHNIQUE OF CRYO-X-RAY TOMOGRAPHY (CRYO-XT)
"STUDY OF FACTORIES VIRAL IN CELLS INFECTED WITH VIRUS "VACCINIA"

AUTHORS

Moreno-Serrano José Antonio. Doctor in Biotechnology and Genetics of Plants and Associated Microorganisms. Professor at the Amazonian Higher Technological Institute. Email: jose79@istam.edu.ec

Morocho Chamba Wagner Roberto. Master in Animal Health. Professor at the Amazonian Higher Technological Institute. Email: rectorado@istam.edu.ec

Jimenez Ávila Luis Antonio. Bachelor of Science Education Computer Mension Educational. teacher of the Institute Superior Technological Amazonian. Email: luis93@istam.edu.ec

INDEX

If you ardently wish to investigate, you should by all means do so. media.

Nothing ought interpose to the desire intense of to dedicate

life to Science. If you have the desire to carry out research scientific [...]: do it!

Hardly some other stuff It will give you so much satisfaction and, above all, such

a sense of accomplishment.

Severe Ochoa

SUMMARY

He V.V. (virus vaccinia) is one of the virus further complex, with a size greater than 300 nm and more than 100 structural proteins. Its assembly involves sequential interactions and important rearrangements of its structural components. In this study, infected cells were selected through microscopy of fluorescence of light and Images were subsequently formed in the X-ray microscope under cryogenic conditions. Tomographic tilt series of X-ray images were used to produce three-dimensional reconstructions showing different organelles cell phones (cores, mitochondria, RE), together with others two types of viral particles related to different stages of virus maturation vaccinia (IV) immature and (MV) particles mature; the essays with witaferin they showed links with actin, which prevents polymerization and elongation of the filaments; causing poorly packaged or aberrant virions, which inhibits the progression of viral infection. The findings demonstrate that the X-ray cryo-tomography is a powerful tool for collecting three-dimensional structural information from frozen, unfixed, unstained whole cells with sufficient resolution to detect different virus particles exhibiting different levels of maturation.

Words clue: virus vaccinia, cryo-tomography of good heavens x, factory viral, actin filaments.

1. INTRODUCCIÓN

Characterizing the complex relationships established between viruses and cells has traditionally provided unique tools to study the virus life cycle and, simultaneously, particular aspects of the cellular systems used by viruses. VV is well known as a very useful expression vector and is now used to design new vaccines against several pathogens (Liu & Moss, 2018; Moss, nineteen ninety six). To the same time, VV has become the focus of cell biologists due to the complex interactions that he virus establishes with the systems cell phones (Hobbs, Osborn, & Nolz, 2018; Ploubidou et al., 2000). A detailed characterization of VV structure and morphogenesis would be of great help for the manipulation of its assembly in vitro and the construction of viral vectors with specific characteristics. Some of the most unknown aspects of the VV morphogenetic pathway are the origin and formation of viral factories (Grossegesse et al., 2018 ; Hollinshead et al., 2001). Viral factories are large cytoplasmic perinuclear areas defined as the centers of VV replication and assembly. The latter occurs in electron-dense masses within the viral factories, known as spotlights of viroplasm (Ding et to the., 2018; Blasco & Moss, 1992). Are structures are formed by he recruitment of items viral, and most likely also cell phones. By mechanisms yet to be defined, the membranous elements adhere to the surface of the VV foci, acquire a curvature and form the crescent. Viral crescents represent the first evidence of assembly of VV, but HE unknown largely as HE form and how they achieve crescents to form immature spherical viruses (IV). There is still considerable controversy about the factories viral, the structures basic and he origin of the crescents VV viral. The use of cryo-electron tomography was an important step towards understanding virus structure (Cyrklaff et al., 2005), assembly (Chichon et al., 2009) and disassembly. (Cyrklaff et al., 2007); But, one of the main limitations of cryo-ET derives from the multiple diffusion penetration of electrons into soft materials (Huang, Li, &

Gao, 2018; Frank, 2006). Even with microscopes of 300 kV of high voltage equipped with energy filters, electron tomographic reconstruction is limited to samples approximately 0.5 µ m thick, precluding direct analysis of most cell types. A complementary approach to overcome these problems is the use of X-ray microscopy. The greater penetrating power of X-rays combined with the progress recent in optics diffractive of good heavens x ha led to the implementation of full-field transmission X-ray microscopes with resolution space in he range of twenty nm (Lau et to the., 2018; Chao et al., 2005). This resolving power, together with the use of soft , thus opening an interesting alternative for he analysis of resolution supramolecular of material biological (Harkiolaki et al., 2018). As in the case of cryo-ET, the combination of X-ray imaging with the preservation of samples at cryogenic temperatures is necessary to recover structural and chemical information close to physiological conditions, as well as to minimize the effect of damage from the radiation on biological samples (Schneider, 1998).The current potential of X-ray cryo-tomography (cryo-XT) for three-dimensional imaging of fully hydrated biological samples is expanding. The technical challenges involved in different aspects of the technique HE find low a wide improvement, such as the use of high-resolution X-ray objectives (zone plates), while solutions to other problems such as precise cryo-sample stages and supports for full-field microscopy are being developed. It has been carried to cape a intense job qualitative for explore the possibilities of cryo-XT with whole cells (Jiménez-Lamana, Szpunar, & Łobinski, 2018; Gu, Etkin, Le-Gros, & Larabell, 2007) that provide morphological descriptions of cellular organelles and segmentation of cellular contents based in properties of absorption differential (Zong et to the., 2018; Parkinson et al., 2008). Cryo-XT studies have played a central role in the characterization of the structure basic, the factories viral and he assembly of VV. This work shows the organization of viral factories and their assembly components in the cytoplasm of a whole cell infected by VV based on cryo-XT and the reconstruction of tomograms by processing the data using specific software.

2. REVISION BIBLIOGRAPHY

2.1. History

Vaccinia virus is a member of the genus Orthopoxvirus of the family Poxviridae. Although No HE know with accuracy, HE believe that HE originated after more than a century of artificial passage of the bovine poxvirus initially used as a vaccine for smallpox in 1798 by Edward Jenner, an English physician. It was in 1930 when HE did evident that the strain that HE used so was different from the one initially used, and from there the Vaccinia virus gained popularity among the medical community as the choice for vaccination against smallpox (Henderson, 1997). Since then Vaccinia has been studied permanently in the laboratory: it was the first animal virus observed under a microscope, grown up in crops cell phones, titled of shape exact, purified and analyzed at a chemical level.Due to its evolution in different parts of the globe, there are several strains: NYCBH (New York City Board of Health) was the strain initially used in vaccination of the smallpox in the state Joined of America; the W.R. (Western Reserve) is a particularly virulent strain derived from NYCBH in the laboratory and that HE uses in the most of the studies and of the essays clinical and the strain Wyeth, that HE uses in vaccines experimental and in essays clinical. Also there is others strains, as the Copenhagen, Lister, IHD-W and IHD-J that HE frequently used for different applications. On the other hand, they have developed strains dimmed as the strain modified Ankara. He genome of some of are strains ha been sequenced (Goebel et to the., 1990; Chan et al., 2000). In the last 30 years the use of Vaccinia virus HE extended beyond its paper in the vaccination of the smallpox, erecting as a useful research tool as a vector for expression of exogenous genes in target cells, So as a half of study of the answer immune of the mammals to viral infection. Its potential is also being explored in therapy against he cancer, mostly of three shapes: as vector for the expression of therapeutic genes specifically in tumors; as a vector for the expression of tumor antigens and/or molecules that stimulate the immune system with he end of develop vaccines against he cancer; and, finally, as an oncolytic virus, specific for highly replicative cells such as tumor cells.

2.2 Virus Vaccinia

2.2.1. Poxvirus

The Poxviridae family is a family of double-stranded DNA (dsDNA) genome and cytoplasmic replication viruses, with very large genomes (130-360 kb in length) that generally encode more than 150 genes (Lefkowitz et al., 2000). Poxviruses are divided into two subfamilies: Entomopoxvirinae, which infect insects, and Chordopoxvirinae, which infect vertebrates (Hughes et al., 2000). Several species of the Chordopoxvirinae subfamily infect humans and domestic animals, especially smallpox virus (VARV), whose only known natural reservoir is humans. This virus is the causative agent of smallpox, a disease that devastated to the population human until his eradication in 1980 after a global vaccination campaign that used a virus closely related called virus vaccinia (VACV) (Bazin, 2000). In 1796, the English doctor Edward Jenner performed the first vaccination in history. with the utilization of a vaccine against the smallpox; made that was a milestone very important in he development of the medicine modern. Besides, is the only vaccine that has eradicated a human disease, and therefore, VACV is the most extensively studied poxvirus.

2.2.2. Genome and Structure of the VACV

VACV is a member of the Chordopoxvirinae subfamily and the Orthopoxvirus (OPV) genus. The VACV genome is a linear, supercoiled dsDNA molecule of 200 kb in length that encodes approximately 200 genes (Lefkowitz et al., 2006). At the ends of the genome there are inverted terminal repeats (ITRs) of size variable characterized by contain sequences repeated in tandem that form a terminal hairpin through the covalent union of both chains (Baroudy & Moss,

1982). The genes of the VACV No they contain introns, are widely separated in the genome, and each gene appears to be controlled by its own promoter transcriptional (Upton et to the., 2006). The genes involved in key functions such as replication, transcription and virion assembly are clustered in the central region of the genome, while the genes involved in the interaction host-virus and in the virulence HE They generally distribute towards the two ends of the genome (Upton et al., 2006). The nomenclature used for to name the different genes of the VACV HE bases in the digestion of its genome by the HindIII enzyme, which gives rise to several fragments of different size. Each gene HE name with a letter in depending on the fragment where it is found after digestion, followed by a number according to the position that occupies saying gene in that fragment, and finally HE Add the letter l either R according to the address of his transcription: left (L) or right (R) (Moss, 2007).VACV viral particles are brick-shaped with slightly rounded edges and dimensions of approximately 360 x 270 x 250 nm (Fig. 2) (Cyrklaff et al., 2006). VACV virions exist in three forms infectious: virions mature (MV), virions wrapped (WV) and extracellular virions (EV) (Smith, 2007). MVs are particles formed by an 8 nm outer layer adjacent to the lipid membrane (Hollinshead et al., 2006). Inside of this layer abroad HE finds a structure porous of 18 nm thick that gives access to a biconcave nucleus, where proteins are located viral structural, he genome of DNA and the enzymes associated to it (Cyrklaff et al., 2005). In turn, the DNA is flanked by lateral bodies that fill the concavities of the nucleus. MVs normally they find each other exclusively within the cells and HE release only by lysis cell phone (Morgan, 1976). The W.V. consist in M.V. which are surrounded by two additional lipid bilayers derived from trans-Golgi cisternae, which contain characteristic viral proteins. WVs are also found inside cells and are precursors to EVs (Schmelz, 1994). EVs consist of WVs that have been released from the cell by fusion of their outermost membrane with the cell's plasma membrane, leaving a M.V. wrapped in a membrane additional. A fraction of EVs is attached to the cell membrane, while others are free in the extracellular medium (Condit et al., 2006). It is believed that the EV are important for the spread of the virus inside of a organism, while MVs are important for long-term stability and transmission of the virus between the guests in he half atmosphere (Moss, 2012).

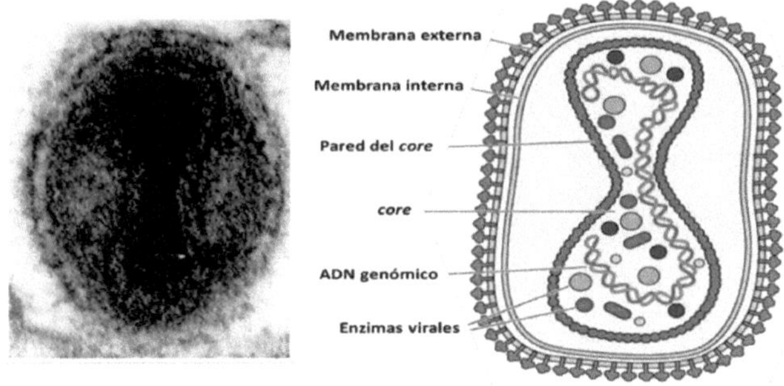

Figure 1. Structure of the VACV. Morphology of the virus Vaccinia. TO the left, electron micrograph of IMV (cross section). The image on the right is a representative schematic of Poxvirus. Translated and adapted from Harrison and collaborators (Cyrklaff et al., 2005).

2.2.3. Cycle Infectious of the VACV

He VACV infects a wide variety of fabrics and his cycle of Replication takes place in the cytoplasm of the infected cell. The viral infection cycle takes place in four steps, namely: entry, disassembly, gene expression and DNA replication, and morphogenesis and release of viral progeny (Fig. 2) (Moss, 2007).

2.2.3.1. Entrance: The entrance of the M.V. HE produces through a fusion direct connection of the viral membrane with the plasma membrane, releasing the nucleus into the cytoplasm, or through a macropinocytosis mechanism, which gives rise to acidified endosomes that also release the nucleus into the cytoplasm. (Moss, 2012). In both processes of fusion, the entrance is assisted by 11 or 12 viral proteins that form the entry fusion complex (EFC). In he case of the EV, that possess a membrane additional abroad, it is produced a mechanism

No fusogenic assisted by proteins viral and cells by which it detaches its additional membrane (Moss, 2012).

2.2.3.2. Disassembly: A time he core viral ha entered in he cytoplasm, is transported through the action of the microtubules toward zones close to core cell phone, where HE form the calls factories viral (Sump et al., 2003). In these places, disassembly occurs, characterized by the loss of viral proteins and lipids and the exposure of the viral genome to the action of exonucleases (Moss, 2012).

2.2.3.3. Gene expression and genome replication:

Replication and transcription of the DNA viral is carried to cape in the factories viral, the which are formed early during infection (Esteban et al., 1984). Transcription of the viral genome is highly regulated and occurs in three stages: early, intermediate and late (Broyles, 2003). Early transcription includes half of the genome and is initiated by the action of proteins and factors of transcription that they came incorporated in he core of the virus (Metz & Esteban, 1972). The mRNAs thus produced encode proteins that are involved in the modulation of the answer antiviral of the guest, the replication of the DNA and the transcription genetics intermediate (Smith, 2007).

The transcription intermediate occurs simultaneously with the replication of the DNA and encodes factors necessary for late transcription (Vos & Stunnenberg, 1988). Finally, late transcription of genes encoding structural proteins, virulence factors and other enzymes takes place (Esteban et al., 1979).

2.2.3.4. Morphogenesis and release of viral progeny: In viral factories HE form virus immature with shape spherical that they mature after to MV due to a prosecution proteolytic of some of the proteins viral and to the condensation of the nucleus (Rodriguez, 1997). Most MVs remain in it cytoplasm until That release after lysis cellular, while a small fraction is transported through the Golgi, where they acquire a second membrane to form the WVs. WVs are transported back by microtubules to the plasma membrane and released as EVs by membrane fusion or projected to adjacent cells via actin tails (Smith, 2007).

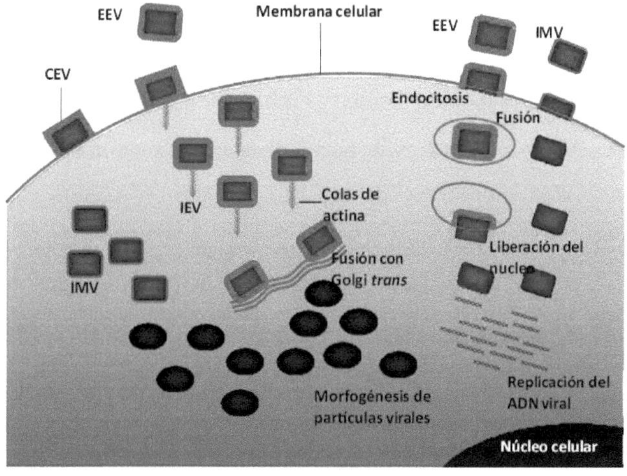

Figure 2. Representation schematic of the cycle of life of the Vaccinia virus .

2.2.4. Interactions Virus-cell Host: Apoptosis and Immune System

Vaccinia virus has developed a series of strategies to avoid the host's immune response. As mentioned above, shortly after entering the cell the virus inhibits a whole series of cellular processes such as the synthesis of DNA, RNA and protein. As a consequence, the production and presentation of molecules stops. of the complex elderly of histocompatibility (MHC), it which carry to poor recognition of infected cells by T lymphocytes. Furthermore, the virus has the ability to directly suppress innate immunity and the Th1 immune response, secreting soluble but truncated receptors for molecules such as IFN- a, py and y (Alcami & Smith, 1995) and IL-18 (Smith, Bryant et al. 2000) that participate in the cellular response to infection viral. The receivers truncated encoded by Vaccinia interfere with the binding of these ligands to their natural receptors on the cell surface, inhibiting his function. Besides, the protein B29R of Vaccinia is a chemokine antagonist (Alcami et al., 1998) and the phosphatase VH1 blocks INF- y signaling (Najarro et al., 2001). Vaccinia has also the ability of inhibit the apoptosis of the host cell. The apoptosis is a mechanism of death cell phone programmed that allows organisms to eliminate infected cells, thus preventing pathogens from replicating in them and spreading to the rest of the organism. To date, four Vaccinia proteins are known to inhibit cell death by apoptosis: SPI-2 (Alcami et al., 1998), E3L and K3L (Fang et al., 2001) and F1L (Wasilenko et al. 2003). Two other proteins, C3L and B5R, inhibit the complement cascade (Kotwal & Moss, 1988). E3L also HE unites to RNA of double chain preventing So activation of P.K.R. (Chang & Jacobs, 1993), and K3L blocks the phosphorylation of eIF-2 a , preventing the antiviral effect of IFN (Beattie et al., 1991). The joint action of these viral proteins allows the virus to infect and replicate in cells without them being able to activate their immune defense mechanisms or programmed cell death.

2.2.5. Interactions Virus-Cell Host: Cycle Cell phone

On the other hand, there is evidence that Vaccinia infection can affect the cycle of infected cells by increasing both the percentage of cells in S phase and the duration of the cell cycle from 24 to 36 hours (Wali & Strayer, 1999). There are other viruses that also affect the cell cycle, such as the virus of Papilloma human and he cytomegalovirus (Jault et to the., nineteen ninety five), that They induce a progression of infected cells from the G1 phase to the S and G2/M phases. It is possible that retention of infected cells in S and G2/M phases facilitates viral gene expression, viral genome replication, and morphogenesis of new virions.

HE ha described also he effect of some poxvirus about the expression of different cellular transcription factors, such as SP1 (Strayer 1993). Specifically he virus Vaccinia seems induce a decrease of the protein levels of p53 and p27 at 12 hours after the start of infection and a decrease in the protein and messenger RNA levels of various positive regulators of the cell cycle such as cyclin B, Cdc2 and Cdk2 (Wali and Strayer 1999). In it that regards to p53, a important regulator of the cellular cycle and of apoptosis, is remarkable he made of that in studies of the effect of infection of two Vaccinia strains (Ankara and WR) in the gene expression of HeLa cells, the transcription of p53 does not seem to be affected (Guerra, Lopez-Fernandez et al. 2003; Guerra, Lopez-Fernandez et al. 2004). Therefore, the effect of the infection of Vaccinia about the levels of protein of p53 HE It must be due to a decrease in the stability of p53. The vast majority of virology studies focus on the biology of the virus life cycle without considering the individual effects that proteins viral they can have about proteins cell phones, and that They may not be necessary for the normal development of the virus life cycle but contribute to viral pathogenesis. That is why we decided to study the effects on cellular proteins of the Vaccinia virus B1 kinase, a protein essential for the development of the virus life cycle.

2.3. cryo-tomography of Good heavens x

2.3.1. History

The history of X-ray cryo-tomography dates back to the development of transmission electron microscopy (TEM) in the 1930s, a technology that revolutionized scientists' ability to study the structure of the cells and the fabrics to a scale subcellular (Dubochet et al., 2018). Without embargo, the techniques conventional of preparation of samples for TEM involved chemical fixation and dehydration, which could introduce artifacts and alterations in the samples biological (Li et to the., 2013). This problem led to the development of cryofixation methods, which allowed the preservation of biological samples in their native state by rapidly freezing them at ultra-low temperatures.One of the fundamental milestones in the development of cryofixation was the introduction of the "vitrification" technique by Jacques Dubochet and collaborators in the 1980s (Mahamid et al., 2016). This method involved ultrafast freezing of biological samples in aqueous solutions, preventing the formation of ice crystals and preserving cellular structure with unprecedented fidelity. Cryofixation opened new odds for the display of samples biological in TEM, allowing the observation of structures cell phones and subcellular in his native state with unprecedented resolution.In the 1990s, researchers began exploring the possibility of combining cryofixation with X-ray tomography to obtain three-dimensional images of frozen biological samples. The integration of cryofixation and X-ray tomography made it possible to overcome the resolution limitations of transmission electron microscopy and obtain high-quality three-dimensional images of biological samples in their native state. In recent years, there have been significant advances in the technology and methodology of X-ray cryo-tomography. For example, new sample preparation methods have been developed that allow better preservation of the structure cell phone and a elderly resolution of image. In addition, improvements have been introduced in detection systems and image reconstruction algorithms, which have led to a significant improvement in the quality of images obtained with this technique.

Today, X-ray cryo-tomography has become an invaluable tool in biomedical research, allowing scientists study the structure and function of the cells and the organelles to subcellular level with unprecedented resolution. As advances in technology and methodology continue, it is expected that this technique will continue to evolve and expand its applications in fields such as cell biology, microbiology and nanotechnology.

2.3.2 Beginning Basics of the Technique:

a. Cryofixation of Samples

Cryofixation is an essential process in cryo-XT, as it ensures the preservation of the structure cell phone in his state native before of be subjected to X-ray radiation. This process, initiated with the freezing technique fast, seeks avoid the training of crystals of ice that could damage biological structures. The vitrification technique, developed by Dubochet and collaborators in the 1980s, has been a crucial innovation in this regard (Dubochet et al., 2018). The vitrification it implies the immersion of the sample in a solution aqueous solution with a high concentration of cryoprotectants, such as ethylene glycol or trehalose, followed of a fast freezing through immersion in nitrogen liquid or through specialized devices. This ultra-rapid freezing process avoid the training of crystals of ice, keeping the cellular structure in a glassy state and thus preserving its morphological and biochemical integrity.Recent studies have shown that vitrification can be optimized through the use of specific cryoprotectants and controlled freezing protocols, which has allowed better preservation of sensitive biological structures, such as macromolecular complexes and cell membranes (Bartesaghi et al., 2018). .

b. Tomography of Good heavens x

Once the sample has been cryofixed, images are captured. using the tomography of good heavens x. This process it implies the rotation of the sample around an axis while acquiring two-dimensional images from multiple angles. Are images two-dimensional it combines computationally for rebuild a image three-dimensional of the sample (McDowall et al., 1983).

The use of X-rays in tomography provides greater penetration compared to conventional electron microscopy, allowing thicker samples to be viewed and high-resolution three-dimensional images to be obtained. Furthermore, current X-ray tomography technology has advanced considerably in terms of image acquisition speed and resolution, which has significantly improved the quality and efficiency of the cryo-XT process. The combination of the cryofixation of samples with the tomography of X-rays ha permitted to the researchers explore the structure three-dimensional of biological samples with unprecedented precision, opening new doors in the understanding of cellular and molecular biology.

2.3.3. Procedure Experimental

a. Preparation of the Sample

The first step in the cryo-XT experimental procedure is the careful preparation of the biological sample. This involves the selection of a suitable sample and its cryofixation to preserve its structure in its native state. Cryofixation is carried out by immersing the sample in a cryoprotectant solution and ultra-rapid freezing to prevent the formation of ice crystals, which could damage cellular structures (Dubochet et al., 2018; Adrian et al., 1984).). During this process, is crucial select he cryoprotectant appropriate and optimize freezing conditions to ensure optimal preservation of the structure cell phone. Besides, the choice of the method of Cryofixation can vary depending on sample type and study objectives, requiring a customized approach for each experiment.

b. Mounting of the Sample

Once the sample has been cryofixed, it is mounted on a suitable support. for his analysis through tomography of good heavens x. This passed involves careful mounting of the sample on a carbon grid or specialized sample holder that allows for manipulation and rotation during sampling. acquisition of images. He mounting accurate of the sample is essential to ensure stability during the tomography process and avoid artifacts in the resulting images (Dubochet et al., 2018).

c. Acquisition of Images

Image acquisition is performed using a transmission electron microscope (TEM) equipped with a ray tomography system. x. The sample mounted HE place in he microscope electronic and HE rotates around an axis while capturing two-dimensional images from multiple angles. These images, known as projections, They are then used to reconstruct a three-dimensional image of the sample using advanced computational algorithms (McDowall et al., 1983; Jiang et al., 2018).

During image acquisition, it is important to optimize the parameters of exposure, as the energy and the intensity of the good heavens x, to get images of high quality with a contrast appropriate and a resolution optimal space. Furthermore, the exposure time must be carefully adjusted for minimize the radiation harmful to the sample, especially in the case of sensitive biological samples.

d. Analysis of Images

Once the three-dimensional image is obtained, the detailed analysis is carried out. of the structures cell phones and subcellular present in the sample. This can to imply the segmentation of structures of interest, the measurement of morphological parameters, such as the size and shape of cells, and the visualization of structures in different planes and sections. Image analysis is usually performed using specialized software of prosecution of images and analysis morphologic. Furthermore, advanced analysis techniques, such as image correlation and three-dimensional reconstruction, can be applied to obtain a deeper understanding of the structure and organization of biological samples (Jiang et al., 2018).

2.3.4. Applications of the cryo-tomography of Good heavens x (Cryo- XT)

2.3.4.1. Visualization of cellular and subcellular structures The Cryo-XT ha permitted to the researchers visualize structures cell phones and subcellular with a resolution without precedents. HE ha used for study the morphology and the organization of organelles cell phones, as mitochondria, reticulum endoplasmic and complex of Golgi, providing information detailed about his function and dynamic (Mahamid et to the., 2016; Leis et to the., 2009).

2.3.4.2. Investigation of Interactions molecular

Cryo-XT has also been used to study molecular interactions within and between cells. By labeling proteins and applying subtomogram techniques, researchers can map the distribution of specific proteins and study the architecture of protein complexes in their cellular context (Bartesaghi et al., 2018; Wan et al., 2020).

2.3.4.3. Study of pathologies human

In the medical field, Cryo-XT has been used to study various human pathologies, such as neurodegenerative diseases, metabolic disorders and cancer. This technique allows the visualization of cellular and subcellular features associated with these diseases, which helps to better understand their underlying mechanisms and develop new therapeutic strategies (Asano et al., 2016; Bridge et al., 2019).

2.3.4.4. Investigation of microorganisms and pathogens

Cryo-XT has been widely used in microbiology to study the structure and function of microorganisms and pathogens. For example, it has allowed the visualization of the ultrastructure of viruses, bacteria and parasites, as well as the characterization of antibiotic resistance mechanisms. and the ID of whites therapeutic potentials (Oikonomou et al., 2016; Liu et al., 2021).

2.3.4.5. Development of nanomaterials biomedical

Besides, the Cryo-XT HE ha applied in he field of the nanotechnology to study biomedical nanomaterials, such as nanoparticles and nanovesicles. This technique provides information detailed about the morphology, distribution and interaction of these materials with cells and the tissues, which is crucial for his development and application in diagnosis and therapy (Souravs et al., 2020; Park et al., 2019).

2.3.5. The Revolution in the Display of Virus: Applications of X-ray Cryo-tomography (Crio-XT)

The display of virus ha been a challenge historical in the microbiology and virology, with fundamental implications for understanding viral biology and develop strategies therapeutics. In this context, the X-ray cryo-tomography (Cryo-XT) has emerged as a revolutionary technique that has transformed our ability to study viruses in their native state with unprecedented resolution. In this essay, we will explore the applications of the Cryo-XT in the display of virus, examining as This technique has contributed to advancing our knowledge of the structure and function of viruses. Cryo-XT is based on the principle of cryofixation of frozen biological samples and X-ray tomography, allowing imaging three-dimensional of high resolution of the virus in his state native. This technique has been applied in a wide range of studies to observe a variety of virus, included he virus of the immunodeficiency human (HIV), the virus of the herpes simple (HSV), the coronavirus and many others. By For example, studies such as that of Bartesaghi et al. (2013) have used Cryo-XT to determine the three-dimensional structure of the HIV envelope glycoprotein, providing crucial information to understand its role in viral infection.Besides, the Cryo-XT ha been fundamental for study the morphology and the organization of virus complex as the coronavirus. Barcena et to the. (2009) performed a study using Crio-XT to examine the structure of the feline hepatitis virus virion, revealing details about the architecture of the viral capsid and its envelope. These findings have not only expanded our understanding of the biology of viruses, but also have practical applications in the development of vaccines and antiviral therapies.The importance of the Cryo-XT in the investigation biomedical is undeniable. Studies such as that of Maurer et al. (2008) have used this technique to investigate the mechanisms of entrance of the virus of the herpes simple in the host cells, providing crucial information for the development of antiviral therapies directed to the inhibition of the entrance viral. Besides, the Visualization of viruses using Crio-XT can provide valuable information for the design of detection and diagnosis strategies for viral diseases, as demonstrated in the study by Zhu et al. (2004) on automatic selection of viral particles. The attenuated vaccinia virus M65 derived from the WR strain (Dallo, Maa, Rodriguez, Rodriguez, & Esteban, 1989) was grown in BSC40 cells and purified by banding on sucrose gradients (Esteban, 1984).

3. METHODOLOGY

3.1. Preparation of Sample for Cryo- ET

Charcoal grids were prepared with the samples of cells infected by V.V. to 12 hpi and 12 hpi + (witaferin), HE They loaded the grids on the copper surface at room temperature for 1 minute. The samples were vitrified, by rapid freezing by immersion in a Leica EM-CPC robot, using liquid ethane cooled with nitrogen (± 178 ºC). The vitrified grids were kept at liquid nitrogen temperature in a storage container before data collection.

3.2. Cryo-Tomography of X- rays

The frozen grids were observed by fluorescent microscopy of light visible in a Leica DMI6000B. The samples Selected samples were transferred to the HZB vibe (photon energy E = 510 eV), with a capacitor type capillary single rebound ellipsoidal glass manufactured by XRADIA (Zeng et al., 2008) with a distance of job of approximately 5 mm and a size of Focal point of approximately 1 μm FWHM (Guttmann et to the., 2009). For cryo-XT used a extension of 2305 times correspondent to a size of 8.68 nm image pixel at 510 eV photon energy.

3.3. Alignment and Reconstruction

Tilt series alignments were performed with the IMOD software package (Kremer, Mastronarde, & McIntosh, 1996), and final reconstructions were performed using the SIRT iterative reconstruction option in TOMO3D (Agulleiro & Fernández, 2011).

4.1. Mapping of the Infection by Virus Vaccinia Through Correlative Soft Light and X-ray Microscopy

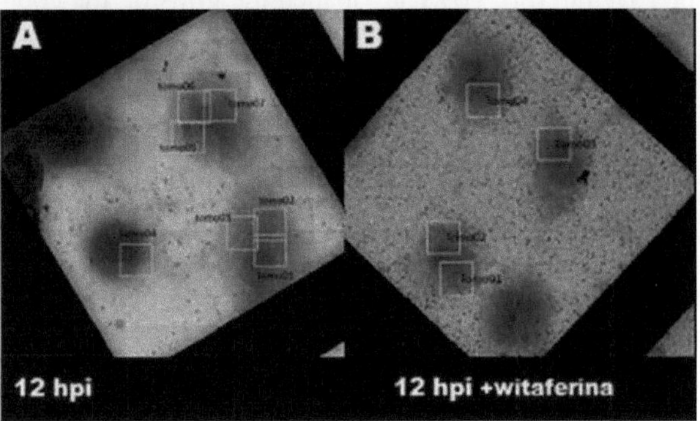

Figure 3. Location of regions of interest and data acquisition in vaccinia virus-infected cells: X-ray cryo-microscopy projection image (zone plate objective with drn = 40 nm, effective pixel size 15.6 nm) of the Labeled cell area: **A)** 12 hpi and **B)** 12 hpi + (witaferin).

The established workflow allowed infected cells to be located using light fluorescent cryo-microscopy. Imaging by cryo-XT was straightforward after focus adjustment for each area. Areas suitable for tomography were selected based on their average density and high contrast. In are areas, the virus HE They found easily as rounded objects. Tilting the sample up to 65° allowed us to obtain well-contrasted images where the viruses were still clearly recognizable. Several tomograms of the same cell were acquired to obtain a tomogram of cells whole during 12 hpi and 12 hpi + (witaferin) (Figure 3).

4.2. Organization of the Factories Viral and their Components Assembly in he Cytoplasm of the Cells Infected with the Virus of the Vaccinia Based in the Tomography of X- rays .

The cells infected by V.V. to 12 hpi showed a displacement of mitochondria and the endoplasmic reticulum in regions distant from the nucleus, generating a cavity that accommodates the virus factory and the viroplasm region. The appearance of discrete cytoplasmic foci of DNA (foci) was observed, also structures filamentous called filaments of actin (F) and a high degree of vacuolization and intracellular vesicles in VV-infected cells. Figure 4 shows the VV assembly area and the formation of the viral factory.

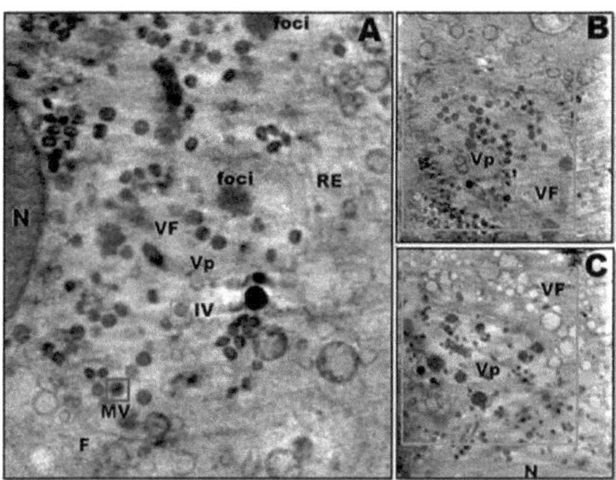

Figure 4. Viral factory at 12 hpi in cells infected with vaccinia virus: **A** , **B** and **C)** viral factory (VF), viroplasm (Vp), discrete cytoplasmic foci (foci), core (N), mitochondria (M), reticle endoplasmic (ER), mature virions (MV) and immature virions (IV).

The studied areas revealed the presence of different associated forms to vaccinia, including particles immature early (Figure 5A) that understand the crescents initials and the called virions immature (IV) (Figure 5B).

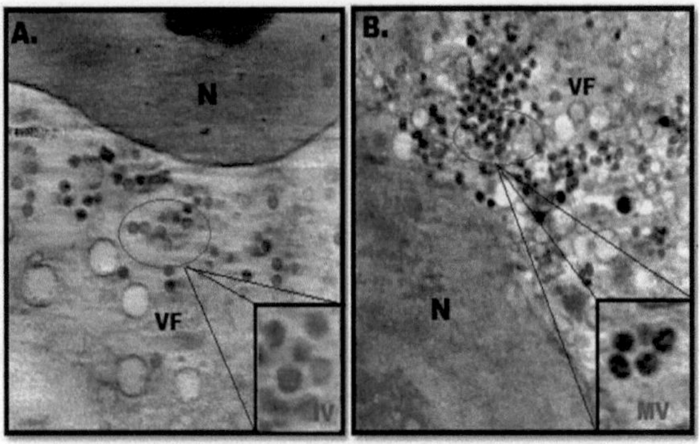

Figure 5. Viral forms recognizable by X-ray cryo-tomography: **A)** virions mature (MV) and **B)** virions immature (IV). He flat virtual a cryo-tomogram of good heavens x acquired with the plate of zone 40nm.

4.3. Training of Filaments of Actin in he Cytoplasm of Cells Infected with the Vaccinia Virus.

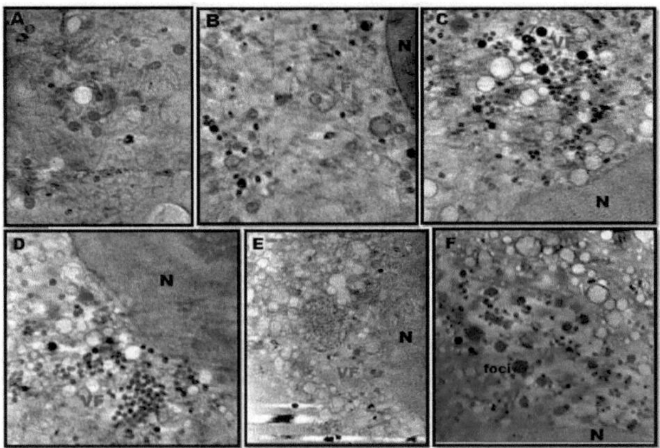

Figure 6. Viral factory at 12 hpi in cells infected with vaccinia virus: **A** , **B** , **c** and **d)** factory viral of control (VF), filaments of actin (F), core (N) and **E** and **F)** viral factory (VF) treated with witaferin, discrete cytoplasmic foci (foci), (N) nucleus.

An assay was performed with BSC40 cells infected with VV, treated and not treated with wiferina to 12 hpi, for determine he effect of the drug on viral replication and virion production. In BSC40 cells treated with witaferin, the presence of intracellular actin filaments disappeared, (Figures 6E and 6F).

5. DISCUSSION

Incorporating cryo-XT microscopy into the analysis of biological materials HE bases in the possibility of penetrate large volumes cell phones (up to 10 μm) for recover information chemistry and structural to Intermediate resolutions between light and electron microscope. The combination of methods based on cryogenic sample preparation and low-temperature image acquisition has led to three-dimensional reconstruction of frozen cells that reveal their internal structures (Chichón et to the., 2012). This study is a tried of obtain information about the possibility to detect relatively small structural features inside of the cytoplasm of cells whole for viral factories and its assembly components in the cytoplasm of VV-infected cells.X-ray cryo-tomograms of VV-infected cells showed several well-preserved cellular organelles, whose structures were easily recognizable as those observed in the microscopy study. by Harkiolaki et to the. (2018). The cells infected by V.V. to 12 hpi showed a displacement of mitochondria and endoplasmic reticulum in regions distant from the nucleus, generating a cavity where the factory is accommodated of virus and the region of viroplasm. HE they observed the appearance of discrete cytoplasmic foci of DNA (phocis), also filamentous structures called actin filaments (F) and a high degree of vacuolization and intracellular vesicles in VV-infected cells. Poxvirus DNA synthesis can generally be detected within 2 h of infection and occurs in the cytoplasm within discrete juxtanuclear sites called viral factories. A virus factory or viral factory can form from a single virion, and the number of factories is proportional to the multiplicity of infection (Moss, 2007). However, the coalescence of the factories individual occurs with frequency with he time (Katsafanas & Moss, 2007). The factories are initially compact and surrounded by endoplasmic reticulum membranes (Tolonen, Doglio, Schleich, & Krijnse-Locker, 2001); A role for membranes in DNA replication has been suggested (Schramm & Krijnse-Locker, 2005). The factory is also the site of transcription and translation of viral mRNAs in addition to virion assembly (Katsafanas &

Moss, 2007).VV assembly and maturation is a complex process that involves drastic rearrangements of structural components. Analysis of such complex intracellular events is technically demanding, and several different hypotheses have been suggested over the years to account for experimental data on morphogenesis obtained primarily by microscopy. electronics. We have done a study of cryo-XT, for obtain a deep insight into intracellular viral particles in the maturation process (stages IV to VM). Furthermore, the use of tomographic procedures allowed us to follow structures and membranes in three dimensions throughout the reconstructed volume, revealing the continuity of the items structural without the issues of interpretation that arise from the superposition derived from conventional two-dimensional analyzes of thin sections.The level of resolution achieved in cryo-XT is such that it has been possible to detect cytoplasmic regions modified by the viral infection (viral factories), as well as two different types of viral particles, the IV and the MV, of particular interest in the context of This studio. Immature virions (IV) with approximately 350 nm in diameter, filled with viroplasmic material and virions mature (MV), that are shapes infectious of approximately 255nm diameter; These results are related to those observed by Chichon et to the. (2009), where he analysis tomographic of the first particle of the assembly VV, he IV, sample particles rounded with a diameter close to 350 nm surrounded by an envelope consisting of a single membrane, this membrane is covered on the convex side by a thick layer (10 nm) of spikes, built by the D13L protein.Such observations lead us to propose that a small number of independent crescents can interact with each other to enclose the volume of an incomplete ellipsoid in the area of the viral factory, reaching the characteristic shape and size of IVs. This proposal is consistent with the observation of type IV particles formed by non-fused crescent-type membranes in mutants lacking the A14L protein, identified as an abundant envelope protein of the first infectious form of VV (Rodriguez et al., 2006). The overlapping regions between crescents neighbors would evolve eventually toward a fusion partial to produce the IV. The partial closure inherent to structure

IV could explain the intrinsic instability of these particles that has prevented their purification until now. Chichon et al. (2012), have tried with limited success different approaches to obtain IV-enriched fractions, only in permeabilized cells after toxin (streptolysin-O) treatment were possible. obtain samples with IV fragile that show contacts extensive with cell membranes.One of the most intriguing aspects in VV morphogenesis is the putative transformation from an immature particle with a single membrane (IVN) to a mature virion (MV) with a single outer membrane and a single other membrane. internal around of the core that contains DNA (Hollinshead et al., 2001). To obtain a vision of this transformation process in this study, cryo-tomography tomograms were analyzed that suggest the presumptive existence of particles with an intermediate appearance IV-MV, these results are related to those observed by Moss (2007), where maturation of IVN toward virions mature HE associate to the presence of particles ranging from rounded to almost brick-shaped, with their internal content clearly different from that of IVN, where the condensed DNA-related material was transformed into a core morphology, in which an incomplete internal membrane was outlining a structure similar to the found in the MVs.The infection viral interferes in the functions normal of a cell to optimize viral replication and virion production. A striking observation of this conversion is the reconfiguration and reorganization of the cellular actin, which affects all stages of the viral life cycle. The extent and degree of reorganization of the cytoskeleton varies between different viral infections, which suggests in this work the possible evolution of viral strategies. Similar results were observed in HSV-1 by Maurer, Sodeik and Grunewald (2008), they detected capsids newly released in close proximity to actin filaments that had apparently reorganized upon viral stimulation to facilitate capsid entry by the elimination of the barrier of the cytoskeleton cortical. The Actin filaments have been studied extensively in vitro (Galkin, Orlova, Vos, Schröder, & Egelman, 2015); without embargo, there is less information available on the precise organization of the actin network in vivo . With recent advances in cryo-XT, and the importance of actin in viral replication and virion production, in this study, an assay was performed with VV-infected BSC40 cells, treated and untreated with wiferin at 12 hpi, to determine the effect of the drug on viral replication and virion production.In BSC40 cells treated with witaferin, it was observed that the presence of filaments of actin intracellular He disappeared, it that suggests that witaferin binds to actin and prevents its polymerization and elongation of the filaments of actin; the replication No HE saw affected, but he assembly and the morphogenesis HE they saw affected in the cells infected with VV, causing poorly packaged or aberrant virions; As a result, the progression of the viral infection was inhibited. It is still unclear how actin filaments interact in vaccinia virus assembly and morphogenesis; Lehmann, sherer, Marks, Pypaert, and Mothes (2005), they studied he effect of vaccinia virus infection, also observing the dramatic reorganization of actin fibers at the cellular level. Likewise, Carlier et al. (1997), in studies with mutant viruses and drugs that inhibit viral morphogenesis, demonstrated that the intracellular enveloped form of vaccinia virus (IEV) is responsible of the nucleation of the filaments of actin. Without embargo, in Another study on baculovirus that also polymerizes actin has shown that filaments of actin his function is stay joined to the virus; still No The function of actin in the vaccinia virus is very well defined, nor is it clear how actin can push membranes and pathogens (Mueller et al., 2014; Small 2015). The replication viral requires the production coordinated, he processing and assembly viral proteins and nucleic acids to produce progeny virions that will be used to infect new cells. Depending on the compartment cell phone in he that HE produces the replication and he assembly, The cytoskeleton has interesting and sometimes surprising functions that are not always well understood.

6. CONCLUSIONS

• X-ray cryo-tomograms of VV-infected cells showed several organelles cell phones good preserved, allowed detect factories viral, like this as two guys different of particles viral, the IV and the M.V. and HE observed the presumptive existence of particles with an intermediate appearance IV-MV.

• It was observed that witaferin possibly binds to actin and prevents its polymerization and elongation, resulting in poorly packaged or aberrant virions, which inhibits the progression of viral infection.

• Cryo-XT opens an interesting possibility to investigate thick samples at the sub-cellular level, as well as the study of the viral life cycle process within infected cells, it is versatile and closes the gap between microscopy of light dynamic and the structures of high resolution obtained by single particle cryo-EM, X-ray crystallography and NMR. The main advantage of cryo-XT is the opportunity to observe processes or molecules in the context close to the native condition at potentially high resolution.

7. REFERENCIAS

Adrian, M., Dubochet, J., Lepault, J., & McDowall, A. W. (1984). Cryo-electron microscopy of viruses. Nature, 308(5954), 32-36.

Alcami, A., & Smith, G. L. (1992). A soluble receptor for interleukin-1β encoded by vaccinia virus: a novel mechanism of virus modulation of the host response to infection. Cell, 71(1), 153-167.

Alcami, A., & Smith, G. L. (1995). Vaccinia, cowpox, and camelpox viruses encode soluble gamma interferon receptors with novel broad species specificity. Journal of virology, 69(8), 4633-4639.

Alcamí, A., Symons, J. A., Collins, P. D., Williams, T. J., & Smith, G. L. (1998). Blockade of chemokine activity by a soluble chemokine binding protein from vaccinia virus. The journal of immunology, 160(2), 624-633.

Agulleiro, J., & Fernandez, J. (2011). Fast tomographic reconstruction on multicore computers. Bioinformatics, 27(4), 582–583. doi: 10.1093/bioinformatics/btq692

Asano, S., Engel, B. D., & Baumeister, W. (2016). In Situ Cryo-electron Tomography: A Post-reductionist Approach to Structural Biology. Journal of Molecular Biology, 428(2), 332–343. https://doi.org/10.1016/j.jmb.2015.09.020

Barcena, M., Oostergetel, G. T., Bartelink, W., Faas, F. G., Verkleij, A., Rottier, P. J., ... & Koster, A. J. (2009). Cryo-electron tomography of mouse hepatitis virus: Insights into the structure of the coronavirion. Proceedings of the National Academy of Sciences, 106(2), 582-587.

Bartesaghi, A., Aguerrebere, C., Falconieri, V., Banerjee, S., Earl, L. A., Zhu, X., & Subramaniam, S. (2018). Atomic resolution cryo-EM structure of β-galactosidase. Structure, 26(6), 848-856.

Baroudy, B., Moss, B. 1982. Sequence homologies of diverse length tandem repetitions near ends of vaccinia virus genome suggest unequal crossing over. Nucleic Acids Research, 10:5673-5679.

Bartesaghi, A., Merk, A., Borgnia, M. J., Milne, J. L., & Subramaniam, S. (2013). Prefusion structure of trimeric HIV-1 envelope glycoprotein determined by cryo-electron microscopy. Nature Structural & Molecular Biology, 20(12), 1352-1357.

Bazin, H. 2000. [Eradication of smallpox, already 20 years ago]. Bull Acad Natl Med, 184:89- 99; discussion 99-104.

Beattie, E., Tartaglia, J., & Paoletti, E. (1991). Vaccinia virus-encoded elF-2? Homolog abrogates the antiviral effect of interferon. Virology, 183(1), 419- 422.

Blasco, R., & Moss, B. (1992). Role of cell-associated enveloped vaccinia virus in cell-to-cell spread. Journal of virology, 66(7):4170–4179. Retrieved from http://jvi.asm.org/c ontent/ 66/7/417 0.full.pdf +ht ml

Bridge, A., Vasishtan, D., & Grigorieff, N. (2019). Cryo-electron tomography of cells: connecting structure and function. Molecular Biology of the Cell, 30(3), 259–266. https://doi.org/10.1091/mbc.E18-10-0661Broyles, S. S. (2003). Vaccinia virus transcription. Journal of General Virology, 84(9), 2293-2303.

Carlier, M. F., Laurent, V., Santolini, J., Melki, R., Didry, D., Xia, G. X., ... & Pantaloni, D. (1997). Actin depolymerizing factor (ADF/cofilin) enhances the rate of filament turnover: implication in actin-based motility. The Journal of Cell Biology, 136(6), 1307-1322. doi: 10.1083/jcb.136.6.1307

Carter, G. C., Rodger, G., Murphy, B. J., Law, M., Krauss, O., Hollinshead, M., & Smith, G. L. (2003). Vaccinia virus cores are transported on microtubules. Journal of General Virology, 84(9), 2443-2458.

Chan, T. A., P. M. Hwang, H. Hermeking, K. W. Kinzler and B. Vogelstein

(2000). "Cooperative effects of genes controlling the G(2)/M checkpoint."
Genes Dev, 14(13): 1584-8.

Chang, H. W., & Jacobs, B. L. (1993). Identification of a conserved motif that is
necessary for binding of the vaccinia virus E3L gene products to double-
stranded RNA. Virology, 194(2), 537-547.

Chao, W., Harteneck, B., Liddle, J., Anderson, E., & Attwood, D. (2005). Soft
X-ray microscopy at a spatial resolution better than 15 nm. Nature, 435(7046),
1210–1213. doi: 10.1038/nature03719

Chichón, F., Rodriguez, M., Pereiro, SAY., Chiappi, M., Perdiguero, B.,
Guttmann, P., & Carrascosa, J. (2012). Cryo X-ray nano-tomography of vaccinia
virus infected cells. Journal of Structural Biology , 177(2), 202–211. doi:
10.1016/j.jsb.2011.12.001

Chichon, F., Rodriguez, M., Risco, C., Fraile-Ramos, A., Fernandez, J., Esteban,
M., & Carrascosa, J. (2009). Membrane remodelling during vaccinia virus
morphogenesis. Biology of the Cell, 101(7), 401-414. doi: 10.1042/BC20080176

Condit, R. C., Moussatche, N., & Traktman, P. (2006). In a nutshell: structure
and assembly of the vaccinia virion. Advances in Virus Research, 66, 31- 124.

Cyrklaff, M., Linaroudis, A., Boicu, M., Chlanda, P., Baumeister, W., Griffiths,
G., & Krijnse-Locker, J. (2007). Whole cell cryo-electron tomography reveals
distinct. PLoS one, 2(5), e420. doi: 10.1371/journal.pone.0000420

Cyrklaff, M., Risco, C., Fernandez, J., Jimenez, M., Esteban, M., Baumeister,
W., & Carrascosa, J. (2005). Cryo-electron tomography of vaccinia virus.
Proceedings of the National Academy of Sciences of the United States of
America, 102(8), 2772-2777. doi: 10.1073/pnas.0409825102

Dallo, S., Maa, J., Rodriguez, J., Rodriguez, D., & Esteban, M. (1989). Humoral
immune response elicited by highly attenuated variants of vaccinia virus and by

an attenuated recombinant expressing HIV-1 envelope protein. Virology, 173(1), 323-329. doi: 10.1016/0042-6822(89)90250-X

Ding, S., Diep, J., Feng, N., Ren, L., Li, B., Ooi, Y. S., & Kuo, C. J. (2018). STAG2 deficiency induces interferon responses via cGAS-STING pathway and restricts virus infection. Nature Communications, 9(1), 1485. doi: 10.1038/s41467-018-03782-z

Dubochet, J., McDowall, A. W., & Elmlund, H. (2018). Cryo-electron microscopy of vitreous sections. The Quarterly Review of Biology, 91(4), 369-407. https://doi.org/10.1086/670067

Esteban, M. (1984). Defective vaccinia virus particles in interferon-treated infected cells. Virology, 133(1), 220-227. doi: 10.1016/0042- 6822(84)90443-4

Esteban, M., Soloski, M., Cabrera, C. V., & Holowczak, J. A. (1979, January). Replication of vaccinia DNA and studies on the structure of the viral chromosome. In Cold Spring Harbor Symposia on Quantitative Biology (Vol. 43, pp. 789-799). Cold Spring Harbor Laboratory Press.

Fang, Z. Y., K. Limbach, J. Tartaglia, J. Hammonds, X. Chen and P. Spearman (2001). "Expression of vaccinia E3L and K3L genes by a novel recombinant canarypox HIV vaccine vector enhances HIV-1 pseudovirion production and inhibits apoptosis in human cells." Virology, 291(2): 272-84.

Frank, J. (2006). Electron Tomography: methods for three-dimensional visualization of structures in the cell. Springer Science & Business Media. Retrieved from https://books.google.es/books?hl=es&lr=&id=LWx6JKQy34AC&oi=fnd&pg =PA1&dq=Frank,+J.+(2006).+Electron+Tomography,+Methods+for+Three dimensional+Visualization+of+Structures+in+the+Cell.+New+York:+Sprin ger.&ots=RnRR_wlrIP&sig=IAQZUD6bquYLuaulwhvr0weoQvo#v=onepag e&q=Frank%2C%20J.%20(2006).%20Electron%20Tomography%2C%20

Methods%20for%20Threedimensional%20Visualization%20of%20Structu
res%20in%20the%20Cell.%20New%20York%3A%20Springer.&f=false

Galkin, V., Orlova, A., Vos, M., Schröder, G., & Egelman, E. (2015). Near-atomic resolution for one state of F-actin. Structure, 23, 173-182. doi: 10.1016/j.str.2014.11.006

Goebel, S. J., G. P. Johnson, M. E. Perkus, S. W. Davis, J. P. Winslow and E. Paoletti (1990). "The complete DNA sequence of vaccinia virus. Virology, 179(1): 247-66, 517-63.

Gu, W., Etkin, L., Le-Gros, M., & Larabell, C. (2007). X-ray tomography of Schizosaccharomyces pombe. Differentiation, 75(6), 529–535. doi: 10.1111/j.1432-0436.2007.00180.x

Guttmann, P., Zeng, X., Feser, M., Heim, S., Yun, W., & Schneider, G. (2009). Ellipsoidal capillary as condenser for the BESSY full-field X-ray microscope. In Journal of Physics: Conference Series (Vol. 186, No. 1, p. 012064). IOP Publishing. Retrieved from http://iopscience.iop.org/article/10.1088/17426596/186/1/012064/meta#art Abst

Grossegesse, M., Doellinger, J., Fritsch, A., Laue, M., Piesker, J., Schaade, L., & Nitsche, A. (2018). Global ubiquitination analysis reveals extensive modification and proteasomal degradation of cowpox virus proteins, but preservation of viral cores. Scientific Reports, 8(1), 1807. doi: 10.1038/s41598-018-20130-9

Harkiolaki, M., Darrow, M. C., Spink, M. C., Kosior, E., Dent, K., & Duke, E. (2018). Cryo-soft X-ray tomography: using soft X-rays to explore the ultrastructure of whole cells. Emerging Topics in Life Sciences, 2(1), 81-92. doi: 10.1042/ETLS20170086

Henderson, D. A. (1997). "Edward Jenner's vaccine. Public Health Report,

112(2): 116-21.

Hobbs, S. J., Osborn, J. F., & Nolz, J. C. (2018). Activation and trafficking of CD8+ T cells during viral skin infection: immunological lessons learned from vaccinia virus. Current Opinion in Virology, 28, 12-19. doi: 10.1016/j.coviro.2017.10.001

Hollinshead, M., Rodger, G., Van Eijl, H., Law, M., Hollinshead, R., Vaux, D., & Smith, G. (2001). Vaccinia virus utilizes microtubules for movement to the cell surface. Journal of Cell Biology, 154(2), 389-402. doi: 10.1083/jcb.200104124

Hollinshead, M., Vanderplasschen, A., Smith, G. L., & Vaux, D. J. (1999). Vaccinia virus intracellular mature virions contain only one lipid membrane. Journal of virology, 73(2), 1503-1517

Huang, X., Li, S., & Gao, S. (2018). Exploring an optimal wavelet-based filter for cryo-ET imaging. Scientific Reports, 8(1), 2582. doi: 10.1038/s41598- 018-20945-6

Hughes, A., Irausquin, S., Friedman, R. 2010. The evolutionary biology of poxviruses. Infection, Genetics and Evolution, 10:50-59

Jault, F. M., Jault, J. M., Ruchti, F., Fortunato, E. A., Clark, C., Corbeil, J.,& Spector, D. H. (1995). Cytomegalovirus infection induces high levels of cyclins, phosphorylated Rb, and p53, leading to cell cycle arrest. Journal of Virology, 69(11), 6697-6704

Jiang, G., Jin, L., Wang, X., Chen, Y., & Wei, D. (2018). In situ cryo-electron tomography: A post-reductionist approach to structural biology. Journal of Molecular Biology, 430(22), 4079-4095.

Jiménez-Lamana, J., Szpunar, J., & Łobinski, R. (2018). New Frontiers of Metallomics: Elemental and Species-Specific Analysis and Imaging of Single Cells. In Metallomics (pp. 245-270). Springer, doi: 10.1007/978-3- 319-90143-

Katsafanas, G., & Moss, B. (2007). Colocalization of transcription and translation within cytoplasmic poxvirus factories coordinates viral expression and subjugates host functions. Cell Host Microbe, 2(4): 221–228. doi: 10.1016/j.chom.2007.08.005

Kotwal, G. J., & Moss, B. (1988). Vaccinia virus encodes a secretory polypeptide structurally related to complement control proteins.Nature, 335(6186), 176-178

Kremer, J., Mastronarde, D., & McIntosh, J. (1996). Computer visualization of three-dimensional image data using IMOD. Journal of Structural Biology, 116(1), 71–76. doi: 10.1006/jsbi.1996.0013

Lau, C., Hunter, M. J., Stewart, A., Perozo, E., & Vandenberg, J. I. (2018). Never at rest: insights into the conformational dynamics of ion channels from cryo-electron microscopy. The Journal of Physiology, 596(7), 1107- 1119.

Lefkowitz, E., Wang C, Upton C. 2006. Poxviruses: past, present and future. Virus Research, 117:105-118.

Lehmann, M., Sherer, N., Marks, C., Pypaert, M., & Mothes, W. (2005). Actin- and myosin-driven movement of viruses along filopodia precedes their entry into cells. The Journal of Cell Biology, 170(2), 317-325. doi: 10.1083/jcb.200503059

Leis, A., Rockel, B., Andrees, L., Baumeister, W., & Visualizing cells at the nanoscale. (2009). Trends in Cell Biology, 19(6), 287-295.

Li, X., Mooney, P., Zheng, S., Booth, C. R., Braunfeld, M. B., Gubbens, S., & Gonen, T. (2013). Electron counting and beam-induced motion correction enable near-atomic-resolution single-particle cryo-EM. Nature Methods, 10(6), 584-590. https://doi.org/10.1038/nmeth.2472

Liu, R., & Moss, B. (2018). Vaccinia Virus C9 Ankyrin Repeat/F-Box Protein Is a Newly Identified Antagonist of the Type I Interferon-Induced Antiviral State. Journal of Virology, 92(9), e00053-18. doi: 10.1128/JVI.00053-18

Liu, Z., Wu, J., Huang, Y., Chen, J., Zhang, H., Jiang, Q., ... & Song, H. (2021). Molecular architecture of the murine norovirus protruding domain and the sialic acid binding site. Journal of Virology, 95(3).

Maurer, U. E., Sodeik, B., & Grünewald, K. (2008). Native 3D intermediates of membrane fusion in herpes simplex virus 1 entry. Proceedings of the National Academy of Sciences, 105(30), 10559-10564. doi: 10.1073/pnas.0801674105

Mahamid, J., Pfeffer, S., Schaffer, M., Villa, E., Danev, R., Cuellar, L.K,& Plitzko, J. M. (2016). Visualizing the molecular sociology at the HeLa cell nuclear periphery. Science, 351(6276), 969-972, https://doi.org/10.1126/science.aad8857

McDowall, A., Chang, J. J., Freeman, R., Lepault, J., Walter, C. A., & Dubochet, J. (1983). Electron microscopy of frozen hydrated sections of vitreous ice and vitrified biological samples. Journal of Microscopy, 131(1), 1-9.

Metz, D. H., & Esteban, M. (1972). Interferon inhibits viral protein synthesis in L cells infected with vaccinia virus. Nature, 238(5364), 385-388

Morgan, C. (1976). Vaccinia virus reexamined: development and release. Virology, 73(1), 43-58

Moss, B. (2012). Poxvirus cell entry: how many proteins does it take?. Viruses, 4(5), 688-707

Moss, B. (2007). Poxviridae: The viruses and their replication (Knipe DM, Howley PM ed.). Philadelphia: Lippincott Williams & Wilkins. p. 2905-2946

Mueller, J., Pfanzelter, J., Winkler, C., Narita, A., Le Clainche, C., Nemethova, M., ... & Schmeiser, C. (2014). Electron tomography and simulation of baculovirus actin comet tails support a tethered filament model of pathogen propulsion. PLoS Biology, 12(1), e1001765. doi: 10.1371/journal.pbio.1001765

Najarro, P., P. Traktman and J. A. Lewis (2001). "Vaccinia virus blocks gamma interferon signal transduction: viral VH1 phosphatase reverses Stat1 activation. Journal Virology, 75(7): 3185-96

Oikonomou, C. M., Jensen, G. J., & Vasishtan, D. (2016). Biomedical applications of cryo-EM of pathogens. Current Opinion in Microbiology, 29, 123-131.

Park, J. S., Kim, D. H., Kim, K., Kim, J. H., Lee, Y. S., Jeong, Y. Y., & Kim, K. (2019). Insights into autophagy machinery by studying Nature Communications, 10(1), 1-12.

Parkinson, D., McDermott, G., Etkin, L., Le-Gros, M., & Larabell, C. (2008). Quantitative 3D imaging of eukaryotic cells using soft X-ray tomography. Journal of Structural Biology, 162(3), 380-386. doi: 10.1016/j.jsb.2008.02.003

Ploubidou, A., Moreau, V., Ashman, K., Reckmann, I., González, C., & Way, M. (2000). Vaccinia virus infection disrupts microtubule organization and centrosome function. The EMBO Journal, 19(15), 3932-3944. doi: 10.1093/emboj/19.15.3932

Rodriguez, J. R., Risco, C., Carrascosa, J. L., Esteban, M., & Rodríguez, D. (1997). Characterization of early stages in vaccinia virus membrane biogenesis: implications of the 21-kilodalton protein and a newly identified 15-kilodalton envelope protein. Journal of Virology, 71(3), 1821-1833

Rodriguez, D., Barcena, M., Mobius, W., Schleich, S., Esteban, M., Geerts, W. Locker, J. (2006). A vaccinia virus lacking A10L: viral core proteins accumulate on structures derived from the endoplasmic reticulum. Cellular Microbiology, 8(3), 427-437. doi: 10.1111/j.1462-5822.2005.00632.x

Schmelz, M., Sodeik, B., Ericsson, M., Wolffe, E. J., Shida, H., Hiller, G., & Griffiths, G. (1994). Assembly of vaccinia virus: the second wrapping cisterna is derived from the trans Golgi network. Journal of Virology, 68(1), 130-147.

Schneider, G. (1998). Cryo X-ray microscopy with high spatial resolution in amplitude and phase contrast. Ultramicroscopy, 75(2), 85–104. doi: 10.1016/S0304-3991(98)000 54-0

Cutter, G., Guttman, P., Home, S., deer leg, S., Eichert, D., & Niemann, B. (2007, January). X-Ray Microscopy at BESSY: From Nano-Tomography to Fs-Imaging. In AIP Conference Proceedings (Vol. 879, No. 1, pp. 1291- 1294). AIP. doi: 10.1063/1.2436300

Schramm, B., & Krijnse-Locker, J. (2005). Cytoplasmic organization of poxvirus DNA replication. Traffic, 6(10), 839–846. doi: 10.1111/j.1600-0854.2005.00324.x

Small, J. (2015). Pushing with actin: from cells to pathogens. Biochemical Society Transactions, 43, 84-91.

Smith, G. 2007. Genus Orthopoxvirus: Vaccinia virus, p 1-45. In Mercer AA, Schmidt A, Weber O (ed), Poxviruses. Birkhäuser Verlag, Basel, Switzerland.

Souravs, M., Patel, K., Singh, R., & Dasgupta, A. (2020). Cryo-EM techniques and its applications in biomedical sciences: Current perspectives and future directions. Journal of Biomolecular Structure and Dynamics, 1-14.

Tolonen, N., Doglio, L., Schleich, S., & Krijnse-Locker, J. (2001). Vaccinia virus DNA replication occurs in endoplasmic reticulum-enclosed cytoplasmic mini-nuclei. Molecular Biology of the Cell, 12(7), 2031-2046. doi: 10.1091/mbc.12.7.2031

Upton, C., Slack, S., Hunter, A., Ehlers, A., Roper, R. 2003. Poxvirus orthologous clusters: toward defining the minimum essential poxvirus genome.

Journal Virology, 77:7590-7600.

Vos, J. C., & Stunnenberg, H. G. (1988). Derepression of a novel class of vaccinia virus genes upon DNA replication. The EMBO journal, 7(11), 3487-3492.

Wan, W., Briggs, J. A., & Cryo-electron tomography and subtomogram averaging. (2020). Methods in Enzymology, 642, 189-215.

Wali, A., & Strayer, D. S. (1999). Infection with vaccinia virus alters regulation of cell cycle progression. DNA and Cell Biology, 18(11), 837-843.

Wasilenko, S. T., Stewart, T. L., Meyers, A. F., & Barry, M. (2003). Vaccinia virus encodes a previously uncharacterized mitochondrial-associated inhibitor of apoptosis. Proceedings of the National Academy of Sciences, 100(24), 14345-14350.

Zeng, X., Duewer, F., Feser, M., Huang, C., Lyon, A., Tkachuk, A., & Yun, W. (2008). Ellipsoidal and parabolic glass capillaries as condensers for X-ray microscopes. Applied optics, 47(13), 2376-2381. doi: 10.1364/AO.47.002376

Zhu, Y., Carragher, B., Glaeser, R. M., Fellmann, D., Bajaj, C., Bern, M., & Potter, C. S. (2004). Automatic particle selection: results of a comparative study. Journal of Structural Biology, 145(1-2), 3-14.

Zong, C., Xu, M., Xu, L. J., Wei, T., Ma, X., Zheng, X. S., ... & Ren, B. (2018). Surface-Enhanced Raman Spectroscopy for Bioanalysis: Reliability and Challenges. Chemical Reviews, 118(10), 4946-4980. doi: 10.1021/acs.chemrev.7b00668

Buy your books fast and straightforward online - at one of world's fastest growing online book stores! Environmentally sound due to Print-on-Demand technologies.

Buy your books online at
www.morebooks.shop

Kaufen Sie Ihre Bücher schnell und unkompliziert online – auf einer der am schnellsten wachsenden Buchhandelsplattformen weltweit! Dank Print-On-Demand umwelt- und ressourcenschonend produziert.

Bücher schneller online kaufen
www.morebooks.shop

Printed by Books on Demand GmbH, Norderstedt / Germany